# Effective Safety Committees

## *A Practical Guide*

D. Rebbitt

**Effective Safety Committees**: A Practical Guide

Copyright © 2018 D Rebbitt

All rights reserved.

ISBN: 172086084X

ISBN-13: 978-1720860846

About the author

Dave Rebbitt is President of Rarebit Consulting providing services across western Canada. He has over 30 years of experience in safety and got his start in 1987/8 with the new WHMIS legislation. He holds of both the CRSP and CHSC designations.

Over his career, he has also become an experienced writer and speaker. He is the most published author, on safety, in Canada. He has published more peer-reviewed safety articles in international journals than any other Canadian safety professional. He also authors several online blogs published in Canada and the USA.

Dave continues to speak at various safety conferences, groups, and companies on management and safety. His articles on pyramids and Hypercompliance, a term he coined, are popular topics.

Dave is also engaged in course development and instruction at the University of Alberta OHS Diploma program. Dave instructs in classroom and online courses ranging from management systems to incident causation.

Dave holds a Master's Degree in Business Administration. His study on the relationship between the number of safety professionals on the fatality rate in several countries remains unique. The research has been referenced internationally, in particular by INSHPO in their review of safety literature and studies.

Dave is also a veteran, having served over 20 years in the Canadian Forces in various capacities. His roles ranged from soldier to technical specialist and manager. His final role was the head of safety for military training facilities in Canada

In his spare time, Dave writes science fiction novels.

## PREFACE

Safety committees have been with us for a long time. Like any endeavor, there are success stories and failures. In many jurisdictions, legislated safety committees are a way for employees to engage with management and solving safety issues in the workplace.

Despite this laudable goal, many committees flounder, becoming less effective than they could be. The key question, in my mind, has always been this. What is the safety committee's purpose? This seems a simple question, and the answer should also be simple. It has been my experience that it is not so easily answered. I hope to do that in this book.

The book is meant to be a guide for committee members and chairpersons. Having the right basics in place can help assure the success of your committee.

Dave Rebbitt

2018

## DEDICATION

This book is dedicated to those that volunteer their time and their efforts to make the workplace safer for others.

Safety committees can be a powerful force for positive organizational culture and a safe workplace. The people who volunteer their time, talent, and perseverance do so in a selfless manner.

This book is dedicated to all those who worked to improve their workplace for their fellow employees.

Got a comment or suggestion?

Send me a message at dave@rarebit.ca

# Table of Contents

## SO, YOU HAVE TO MAKE A SAFETY COMMITTEE?

Many people who read this book will do so because they were directed to make a safety committee.

It seems a daunting task. However, rest assured you are not alone and not the first person to find yourself in this particular dilemma.

A safety committee is simply a group of people that provide direct feedback on the operation of the safety management system, or safety program, in the company.

The information you find in this book will help you set up a safety committee. First, you must determine what the company expects the committee to do. Only after you've done this, would you be able to start to recruit members.

Recruiting members can be very difficult, and so you must have a very good idea of what the safety committee will do. If you are the one setting up this committee, you should draft the terms of reference so that the committee can at least have something to talk about it their first meeting.

Do not think that people can simply be appointed. Even appointees should agree to be appointed. Most people asked to join a safety committee, especially one that is just being formed, will last for the same basic questions.

1. What is the committee's purpose?
2. What am I expected to do?
3. How much time do I have to commit to this?
4. Will I get any training?
5. How long am I expected to serve on this committee?

You must be able to answer these questions with confidence. Once you understand what the committees function will be within the company, or within the facility, you must identify what resources will be available to the committee. You should expect that training will be funded for the co-chairs as a minimum.

Your plan to form a safety committee begins with validating its purpose and verifying appropriate resources have been allocated for the committee.

Below you'll find a simple checklist that is related to the specific sections of the book. Although the checklist appears simple, it may take months to get everything in place before the committee can have its first meeting.

After that, it may take more months before the committee is operating normally. The keys to success are structure, training, and a phased approach.

If your company has a health and safety person, or department, do not hesitate to ask for their assistance in setting up the committee.

Here are the 10, not so simple, steps to set up your safety committee.

1. **Determine the committee's purpose.** This means understanding what the committee is expected to do.

   There are many aspects of the committee can become involved in, and you may wish to take a phased approach in easing the committee and to being fully integrated into monitoring the health and safety management system.

   At this point, you also want to determine what the structure of the committee will look like.

2. **Verify that sufficient resources have been allocated.** Members volunteering for the committee, or being appointed to the committee, will expect that they receive some training.

   You may choose to do an in-house training session by bringing in a consultant or training specialist on committees. This may be a good time to build the committee's terms of reference.

3. **Recruit members.** This means speaking to people in order that they might agree to be appointed. It also means marketing the committee before asking for volunteers.

   In a union workplace, you would meet with the union to determine how they would like to fill the committee.

4. **Get key members trained.** Key members, such as Co-chairpersons should be trained. This would, ideally, happen before the first meeting of the committee.

   This may not always be possible, and so the idea of having a workshop as the first meeting is much more important.

5. **Draft a terms of reference.** The committee needs a terms of reference, and someone must draft it. This can be done as a workshop and training session for the committee.

   However, it can also be drafted by a single person for review and revision at the committee's first meeting.

6. **Schedule the first meeting.** Scheduling the first meeting means clearly communicating the meeting and building an agenda. This process is covered under "meetings" in the book.

7. **Finalize terms of reference.** Before the committee starts to conduct any business, it is important that the committee finalizes its terms of reference so that everyone on the committee would clearly understand how the committee was meant to operate.

8. **Bring the committee up to speed.** In the first few meetings, the committee cannot take on all its duties. These meeting should concentrate on making the committee members familiar with how they would be reviewing or participating in incident investigations, how they would be

interfacing with the company in terms of handling concerns, and how the meetings will be structured.

Having a few meetings to take on the committee's duties in bite-size chunks will help avoid committees from becoming overwhelmed.

9. **Review the terms of reference.** Once the committee has met a few times, it may be necessary to review the terms of reference to ensure that the committee feels they are reflective of what the committee wishes to accomplish.

It is at this point that the terms of reference may be revised to reflect the realities of the workplace that the committee operates represents.

10. **The committee is in normal operation.**

## WHAT IS A SAFETY COMMITTEE'S PURPOSE?

A health and safety committee is a joint effort between employer and employees to improve workplace safety. However, each member of the committee would probably have a different answer to this simple question. Each company has its own vision of what the safety committee should be for.

At the outset, it is important to describe the committee's role and how they will fill that role within the company. Legislation often indicates that committees would be involved in certain activities like inspections or incident investigations. It sometimes details, to some extent, how that involvement would take place. Often, it is left to the committee to define their level of involvement and how that will work.

The committee does this through its terms of reference. This is a document that may be drafted with the assistance of safety personnel, or by the committee itself in its first few meetings.

In answering the question, about what the safety committee's purpose, is done through the drafting, approval, and revision of robust terms of reference. This is the document that defines how the committee operates and what it will do. This also allows the committee to measure its performance and success.

There are some things to consider when defining the committee's purpose. Here are some questions to get you started.

1.  What does the employer expect from the committee?

2.  Who will the committee represent?

3.  How will the committee be involved in:
    a.  Setting safety strategy?
    b.  Setting safety goals and targets?

4.  What things will the committee focus on improving?

5. How will the committee integrate with existing functions to provide oversight in such things like:
   a. Incident investigations?
   b. Inspections?
   c. Workplace monitoring or testing?

6. How will the committee get, or encourage input from the people in the workplace(s) it represents?

7. What are the current areas of focus in the company when it comes to safety?

8. What are the current safety targets?

9. What are the current new initiatives or programs in the safety management system?

10. What do the workers expect from the committee?

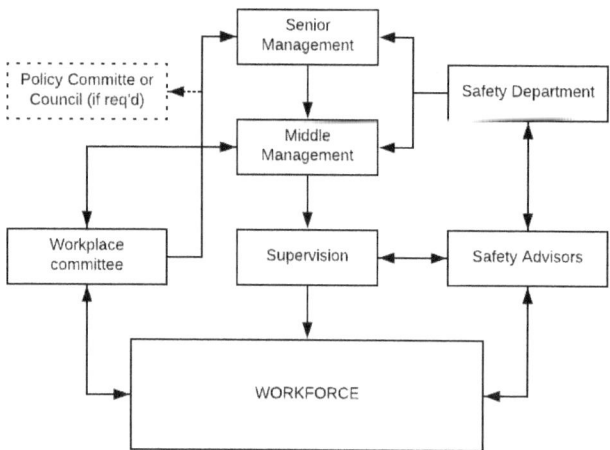

**Figure 1:** *The health and safety committee forms an integral part of the company's safety system. It allows direct communication of concerns at the workforce level to the senior management level and into the company's health and safety department, or function*

## TERMS OF REFERENCE

When forming a committee, the primary task is to develop terms of reference for the committee. The terms of reference will describe all aspects of committee operation and should be ratified by the committee and approved by the employer. The terms of reference often become a document that is included in the safety management system.

The terms of reference can encompass a wide variety of topics and areas.

### General Terms of Reference

The actual terms of reference documents for committees vary widely. A Terms of Reference (ToR) is the guide that helps members understand how the committee functions, and also helps explain to those in the workforce what it is the committee actually does.

Many things merit consideration for inclusion in the terms of reference. These topics are covered in more detail within this section.

Areas that should be addressed in the terms of reference include:

1. The purpose of the committee;
2. What areas the committee represents;
3. How the committee fits into the overall company structure;
4. How many members should be on the committee;
5. How many members of the committee will be management, and how many will be employees, or labor;
6. Duties of members;
7. Duties of Co-chairpersons;
8. Training required for committee members or Co-chairpersons;
9. The role of the safety department or safety personnel
10. How members will be appointed or elected

11. Length of member's term on the committee;
12. What constitutes a quorum;
13. How often the committee will meet;
14. How the committee's records will be kept; and
15. How, and under what circumstances, the committee will make formal recommendations to the employer.

This list is by no means complete, but it gives a good general idea of what should be considered when writing a committee terms of reference.

**Purpose of committee**

The terms of reference should identify the purpose of the committee and how it will conduct its business. It should define what the committee should be involved then and what its regular activity should be.

The purpose of the committee should be like a mission statement. It should be a short way of communicating the mission of the committee. For example:

The area safety committee exists to work collaboratively with the employer to maintain and improve workplace safety. It will accomplish this by actively seeking worker input on safety matters and responding to concerns identified. The committee will provide oversight of key safety processes to ensure they are effective in maintaining safety.

The purpose of the committee may also include the duties or responsibilities of the committee.

For example, some duties that the committee may undertake are:

- to work cooperatively with the company to resolve issues related to safety;

- to assist in identifying and correcting workplace hazards;

- to identify, evaluate and participate in the resolution of matters about health and safety in the workplace to appropriate managers;

- to encourage adequate education and training programs so that all employees are knowledgeable in their rights, restrictions, responsibilities under the legislation;

- to assist and consult in the development and maintenance of the health and safety program of the organization;

- to promote health and safety in the workplace;

- to make recommendations to the employer and the workers for the improvement of the health and safety of workers;

- to recommend the establishment, maintenance and monitoring of programs, measures and procedures respecting the health or safety of workers;

- to obtain information from the employer respecting the identification of potential or existing hazards of materials, processes or equipment;

- to obtain information from the employer concerning the conducting or taking of tests in the workplace for occupational health and safety; and

- to be consulted about testing conducted in or about the workplace and receive such information as required to determine the validity of the testing.

## COMMITTEE, OR COMMITTEES?

In larger companies with more than one in a geographical location, there may be a requirement to have more than one committee. If there is more than one committee, the activities of the committee should be coordinated.

In larger companies, there may be a higher-level policy committee that includes members from the other committees. In the medium-sized companies, it may fall to a single person to collate the committee activities for senior management.

Wherever there is more than one committee, there is a need to ensure some cross-communication between the committees to avoid redundant efforts and to share solutions.

In some cases, the company's safety department may have a member assigned to each committee to provide assistance and facilitate communication between committees so that strategies, solutions, and common issues can be shared.

This will be dealt with later in the book.

## COMMITTEE STRUCTURE

It is important to define the structure of the committee at the outset. The structure of the committee must be at least 50% employees. The structure is meant to address just how many people should be on the committee.

Most jurisdictions require a minimum of four people. It is suggested that any workable committee should not have more than ten members or so. Larger groups make for longer meetings and less productive ones. In large workplaces, it may be necessary to have more than one committee and the committee system.

### 50-50 approach
Committee makeup tends to follow the 50-50 rule in that there is 50% membership from the worker/employee side and 50% from management. This is the simplest approach and initially the most common.

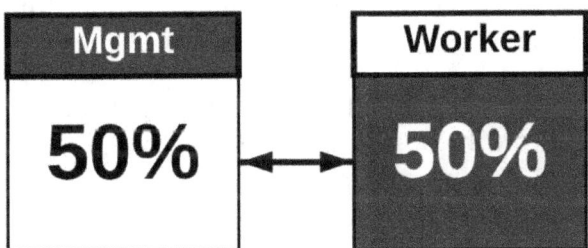

### 60-40 approach
Most requirements for safety committees require at least 50% of the committee members represent workers, but there is no upper limit.

For example, the committee structure may require specific representation from different areas or departments. The committee structure may require that discipline such as maintenance or operations are represented. The structure may also concentrate on geographical locations to make sure those are represented. In some cases, the committee structure may focus on specific departments to ensure that those departments have a voice within the committee.

In most cases, having good representation from the work areas at the worker level on the committees means that healthy committees have more than 50% worker members.

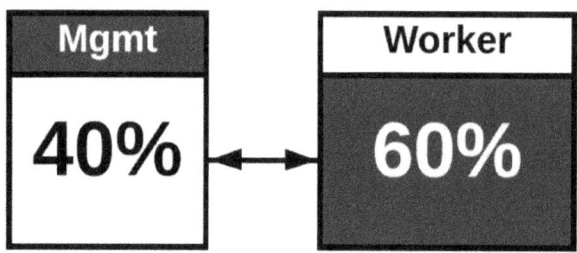

The 60-40 approach also makes it easier for the committee to establish a quorum for meetings. All areas can have worker representation with some management representation. With the 60-40 approach, a ten-member committee would have four management members and six worker members.

This would allow the committee to meet even with two worker members absent. In the 50-50 approach, the committee would not be able to meet with one worker member was absent.

Committees that are structured under the 60-40 principle tend not to miss meetings and can be more effective since there is more opportunity for input from the worker level, which is important for the committee to be effective.

**Membership**

The membership of the committee is often dictated by legislation as being at least 50% employees or labor. However, it is important that all areas of the business at the committee's location be adequately represented.

If the workplaces are unionized, recruiting members often falls to the union for the employee side of the committee. Having a union of point members to the committee, or have them elected at a union meeting, ensures that there is adequate representation in the view of the union on the committee.

It is important to involve any bargaining agent in the selection of committee members as they will be representing the interests of employees, and therefore, the union. When forming a committee, a discussion should be held with senior officials of the union to ensure that the number of representatives and the areas represented is clear. This will also help establish the participation of the union in providing representatives for the committee.

For non-unionized workplaces, participation in the committee is generally on a voluntary basis. This makes the establishment of terms of reference even more important as employees would only volunteer to participate in a venture they saw as effective.

Volunteers usually join the committee because they have an issue that they would like to see resolved. To ensure that such volunteers do not leave the committee once they see their issue resolved, it is important to show that the committee does do valuable work.

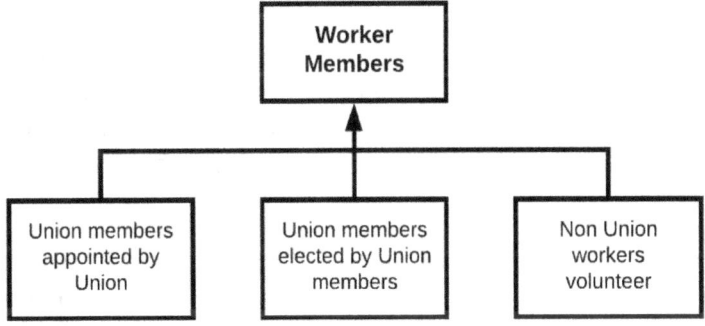

Soliciting for volunteers when forming a committee is often a straightforward enterprise. It is often wise to enlist the support of senior management in putting an announcement that a committee is being formed and that volunteers are being sought.

Management, or employer committee representatives, are usually appointed by senior management. It is important that at least one employer representative be a manager, or senior manager, that has some decision-making authority and resources.

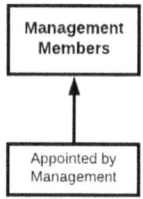

The senior manager who is making such appointment may change from time to time, but it should be understood who will do this.

Senior management may require candidates to be submitted to them for consideration prior to the final appointment of management candidates. In such cases, a summary of each candidate, and how they may benefit the company, may be prepared for consideration.

## Appointment of members

The terms of reference should describe how members of the committee will be appointed. As discussed under the heading of "Membership," the terms of reference must outline how members are appointed. This is important as members may leave the committee from time to time and they must be replaced.

The terms of reference you describe how management will appoint employer representatives. Companies may leave that decision to a single person or several people. In nonunionized companies, it is sometimes a case that even the employee representatives would be appointed.

A clear process for the appointment, election, or volunteering of committee members helps ensure that there are not long periods of time where there are vacancies on the committee. This also helps avoid the cancellation of meetings where a quorum cannot be gained.

## Lengths of member terms

It is important to define how long an appointed, or elected member may serve on the committee. Setting this term at two years is popular. This allows members to address issues that they see in the workplace but also allows for wider participation in the committee. Many committees allow members to be reappointed, or re-elected, only once.

In some cases, exceptions are made for the chairpersons as management may want the same person to chair the committee until they feel a change is needed. In other cases, a chairperson term is viewed as a separate term from that as a member.

Overall, a period of 2 years on the committee as a chairperson, or a member, is standard. The flowchart below represents the common flow for membership but is only an example of how this may be structured.

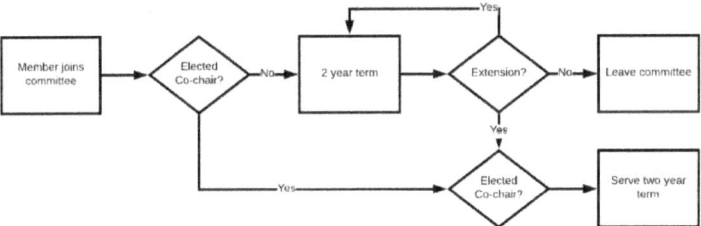

**Figure 2:** *Members length of terms*

## Election/appointment of chairpersons

Chairpersons are the ones who chair the committee meetings. This is a responsibility shared between the two chairpersons as each takes a turn in chairing the meeting.

It is common for the employee chairperson to be elected from the employee members of the committee. This is often done by secret ballot in an informal setting, usually at a committee meeting. In unionized workplaces, the union may appoint a chairperson rather than have an elected one.

In some cases, the union would do hold the election of the employee, or worker, chairperson at a union meeting. The important point is that it is the members of the committee representing employees would elect their chairperson unless the union has some other method they prefer.

The employer chairperson is normally appointed by the employer. However, the employer chairperson may also be elected by the employer members of the committee if that is what is desired.

The co-chairpersons are the linchpins of a functioning committee. They each would take turns chairing the meeting and being responsible for the meeting they chair in that they would finalize the agenda and ensure that the appropriate communications are made to members and the workforce.

### Duties of members

Members of the committee have specific duties. This would include co-chairpersons as well. The terms of reference should define expectations for those joining the committee so that they understand their role and their responsibilities.

Some example duties of members.

Committee members shall:

- Attend all committee meetings;
- Notify the presiding chairperson if they cannot attend a meeting;
- Solicit input from their Regions/Departments as appropriate; and

- Review written materials, and previous minutes before meetings.

## Duties of chairpersons

Chairpersons have their own specific duties. Generally, it falls to the chairperson to ensure that meetings are properly scheduled, planned, and managed.

Chairpersons are an important part of the safety committee as they normally have the most training in health and safety. Additionally, chairpersons must ensure they retain control of the meeting to ensure that the committee does not lose focus and purpose.

Some examples for chairpersons appear below.

Presiding Chairperson shall:

- Schedule meetings;
- Prepare an agenda;
- Invite specialists or others, as required;
- Preside over the meeting and guide it as per the agenda; and
- Review and sign off on the minutes.

## Duties of the Secretary

Secretaries may, or may not, be part of the safety committee. The employer may choose to provide administrative support for the committee, and the secretary normally would have at least these basic duties:

- attend meetings and take the minutes;
- gather agenda items on behalf of the chairperson;
- have the co-chairpersons sign the minutes of the committee;

- distribute the minutes to the workplace and members of the committee;
- distribute the meeting agenda to committee members before each meeting; and
- act as a liaison for visitors to the safety committee meetings

## Meetings

The terms of reference for the committee should specify all aspects of meetings. Many committees will establish a certain day that they would meet.

A committee that meets monthly may meet on the 3rd Thursday of every month, for example. The terms of reference would include much information about meetings. The terms of reference should address things about meetings such as:

1. How often will the committee meet;
2. How will members be notified of the committee meetings;
3. How will the employees be notified that the committee is about to meet so they can make their representative aware of any concerns they might have;
4. How the agenda for the meeting will be communicated;
5. How any information that will be covered in the meeting will be communicated to the members; and
6. The duration of meetings.

## Quorum

To conduct business, a committee must have a quorum. Is important the terms of reference define what a quorum is to avoid confusion. A quorum normally means that at least half of those present are employee representatives and that at least half of the committee is present.

In some cases, the committee may decide to attach a number to a quorum. If the committee has a dozen members, for example, the terms of reference may state that a quorum must be at least six members, where at least 3 of them are employee representatives.

It is a common practice to insist that at least half of the committee should be present to establish a quorum. This avoids decisions being made for the majority by the minority.

## Concerns or complaints

The committee will receive from time to time concerns or complaints related to workplace health and safety. It is important that the committee determine how it will handle these concerns or complaints.

Complaints or concerns should certainly be logged, but some investigation should be done to ensure that the complaints/concerns are validated.

The committee should determine how it should receive such concerns or complaints. Requiring employees to fill out a form would probably make people much more reluctant to come forward.

Only concerns/complaints that were deemed valid by the committee should be recorded in the minutes.

## Training

Training for committee members is usually not well defined. While there are normally provisions for committee training, it is important that the co-chairpersons be trained.

The content of that training is usually open to definition by the employer. Some suggestions for training include:

1. OHS Legislation (awareness or more in-depth course);
2. Legislation governing committees;

3. Safety committee duties and responsibilities;
4. Safety programs or systems;
5. Hazard identification, assessment, and control;
6. Risk management;
7. Workplace inspections;
8. Incident investigation and cause analysis; and
9. Running effective meetings/Effective communications.

While this training may require attending several courses, training members of the committee is an investment in their success in helping to improve workplace safety. Training in specialty areas such as fall protection, respiratory protection, or confined space entry may be needed in addition to the basics to help committee members understand workplace issues.

## Payment for your attendance at meetings

Payment for the attendance of a safety committee meeting is often straightforward. However, in a unionized workplace, this may not be clear.

The safety committee is often considered union business and so may be addressed within the collective agreement as such. Often unionized members of the committee are paid their standard hourly rate for attendance to meetings. However, where a new safety committee is being set up, it is important to address this before the first meeting to assure employees that they will be paid for this important function.

## Integration into safety functions

Some key functions of the safety committee may be that they would participate in incident investigations, inspections, and even work refusals. The terms of reference should address how committee members would be involved in the specific functions as these would be done outside of committee meetings.

For example, the committee may be satisfied with simply reviewing completed incident investigation reports. In some jurisdictions, there is a requirement to involve a safety committee member in these functions more directly. Where there is a committee, it is almost always required, and prudent, to involve them in the investigation of serious incidents or major losses.

Most committees are expected to do inspections of the worksite and regular intervals that are specified by the committee. Some committees dispatch a team to conduct an inspection immediately after their meeting, or immediately before. These inspections can be monthly or quarterly. The committee members may choose to inspect in conjunction with other employees as specified in the committee terms of reference or the company health and safety manual.

Work refusals are rare. These can be high-stress events for those involved, and the involvement of a safety committee member as an impartial party can be helpful. If safety committee members are to be involved in work refusals or stoppages, it is important to define how that interaction would work within the terms of reference.

Finally, the safety system within the company should be updated to include the involvement of the committee in the key functions.

The committee terms of reference are part of the health and safety management system and may be included in the health and safety manual unless there is more than one committee. In cases where there is more than one committee, a general standard governing committees should be included in the manual.

Even after involvement in the key functions is defined, the committee would still define his involvement in other functions such as training development, or emergency response planning as it progresses.

## Record-keeping

It is important that the committee keep records of its agendas, minutes, and reports that the committee examines. Normally some administrative support would be assigned to the committee by management to ensure that adequate minutes are taken during the meeting.

In some cases, co-chairpersons may take the minutes. That it is often difficult for them, as they must control the meeting and participate in it. This is often not an effective use of the co-chairperson's time during a meeting.

Records are often Electronically at a central location. This allows easy access for any committee member to access the electronic records and review them at their leisure.

The terms of reference should describe how the minutes will be distributed or made available to employees.

## RECOMMENDATIONS

From time to time, it may be necessary for the committee to make formal recommendations to the employer. This is an important function that the committee may serve to assist in resolving health and safety issues that arise.

The terms of reference should define how these recommendations should be structured and how they will be communicated to the employer. A subcommittee often does this.

Recommendations should give a background of the issue so that the reader could understand why this issue exists, and why it needs to be addressed. It is also important that the committee help the employer understand the risk related to the issue that is being raised. This will assist the employer in understanding the risk posed by the issue and reacting appropriately.

The committee should make clear recommendations and include the rationale for the recommendations. It is important that the thought process of the committee and the options presented by it show that they are carefully thought out and reasonable.

Recommendations should include a series of options rather than a simple request to do a specific thing. It should be up to the employer to determine how they would act in response to a recommendation. Presenting a series of options would reduce the time required to resolve an issue while acknowledging that the employer is fully responsible for the outcomes.

There are some important elements of recommendations and some minimal information that must be included. A recommendation should be entirely self-contained. That means it should, whenever possible, contain all the information required to start a decision-making process.

As a minimum, the recommendations should contain:

- **Background.** The first part of the recommendation should contain a summary of information or events that have led to the recommendation. This is critical so that a reader may understand the current situation and see the issue as the committee sees it. Information regarding testing, incidents, observations, or inspections can be included to provide depth to the perspective and help the reader understand why the recommendation or recommendations are being made.

- **Discussion.** This is the discussion of information presented and how this issue is affecting, or may affect, the workplace. It also explains why the step is being taken to make formal recommendations. Here it is often wise to discuss risk and risk ratings to establish an urgency for action.

- **Recommendation.** Recommendations must be clear and in a single sentence where possible. Avoid phrases like "the company should..." or "A program needs to be implemented..." A recommendation needs to be clearly identified and laid out. For example:

  **Recommendation 1.** Conduct third-party certification of rigging equipment.

  The recommendations, when clearly stated, are easier to understand from all sides and describe the specific action. Where possible the recommendations should be listed in order of priority, or in the order they should be completed.

- **Options.** Options are about how a recommendation is met. It is often wise to consider if the action must be undertaken within a few months or a phased approach can work over a longer period of time. It is also important to remember that doing nothing or declining the recommendation remains an option.

- **Rationale.** The rationale, or reasoning, for each recommendation, needs to be explained so that the reader and the company may understand why this recommendation is being made and what the benefits to the company may be.

- **Cost.** Most recommendations have a cost attached, and so the costs should be researched to ensure an understanding of the cost attached to any action. In some cases, the rationale for a recommendation may be driven by cost. The cost for each recommendation, or option should be identified.

- **Summary.** This ties the recommendation together by briefly reviewing the issues and the benefits of implementing recommendations. It can also serve to add some urgency to the recommendation where appropriate.

By providing a complete, and well-reasoned, recommendations, the committee can present management with the information it needs to make a decision and ensure that management is aware of that effects of that decision.

## THE ROLE OF THE SAFETY PERSON/DEPARTMENT

When the committee is being formed, it is often the safety person or safety department, that quarterbacks this activity. Safety personnel can be of great assistance in defining the safety committee's role and helping get the committee up and running.

The role of the safety committee is often to review and comment on the effectiveness of the safety management system or its elements. As such, the safety person or safety department cannot really be members of the safety committee since they would, in effect, be assessing their own performance.

It is often tempting to appoint the safety person, or safety manager, to head the committee or act as one of its members. This is often what leads to ineffective committees.

The role of safety personnel in committee should be ex officio, meaning that they might attend meetings but have no vote, or stake, in the proceedings. Safety personnel are best used as a resource for the committee. They can be used to provide information and assist in the interpretation of information.

The committee organizational chart would often depict the safety department or persons as advisors to the committee as denoted by the dotted line report from the side rather than below.

Using in-house safety personnel for their knowledge and skills will help make the committee effective. Making them part of the committee, or chairperson of the committee, often leads to an ineffective committee.

Safety personnel can assist the committee in interpreting reports and guiding them through the safety processes of the company.

In some cases, it may be beneficial for a safety person to provide a short training session to the committee on specific aspects of the safety system, or program. Such sessions would be for the information of committee members rather than to qualify them in a specific area.

This can be of great assistance in helping all committee members understand the safety system, or program, at the company.

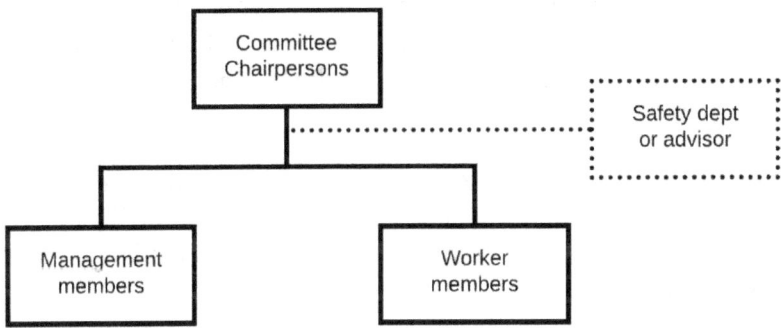

***Figure 3:*** *The safety person or department would act in an "ex officio," or unofficial capacity acting as a resource to the committee, providing assistance as required.*

## MULTI-COMMITTEE WORKPLACES

As identified in the section on terms of reference, some workplaces may have more than one committee by virtue of the size of the organization or its geographical diversity. To ensure participation from all areas of the company in these cases, it is often necessary to have more than one committee, and a means to coordinate their efforts.

In some cases, the geographical worksite may have made different functions on it and so will the decision could be made by senior management to have more than one safety committee to ensure that concerns from all areas had an outlet and that broader participation was encouraged.

Often multi-committee workplaces go beyond the minimum legislative requirement and so it is up to the employer to ensure that committees all have a similar makeup and steward to the required legislation.

This would mean that each committee would have the same basic structure and terms of reference. Ensuring a commonality to the function and duties of committees will build the foundation for a successful committee structure.

With various committees at a single geographical location, it may be necessary to have an oversight body such as a safety council to coordinate the activities of the committees and to deal with problems that may affect the entire company or should be elevated beyond the capability of a single committee.

### Committee structure

The committee structure may see two, or even more, levels of committees within the employer structure. The need for more than one committee is something that only usually arises for large and very large companies. Some examples of committee structures appear below.

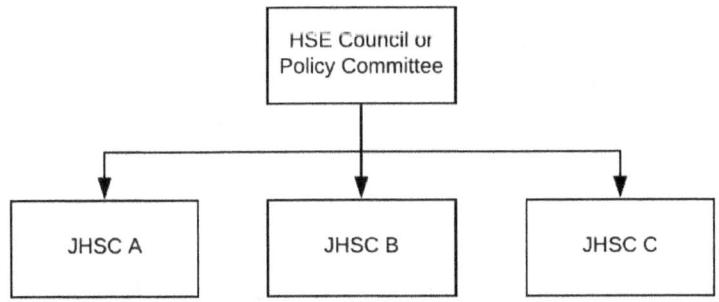

***Figure 4:*** *Basic committee structure with three workplace committees and a company level council or policy committee.*

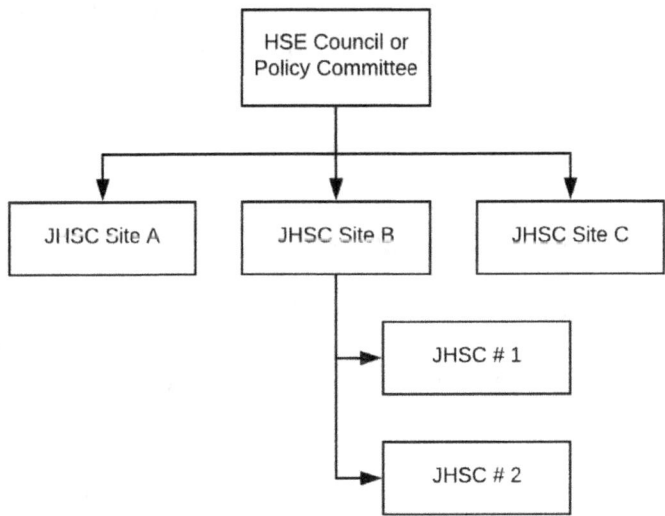

***Figure 5:*** *Committee structure where Site B, being a larger site, has several safety committees. Each site would have a safety committee, but site B would be more than one committee to represent and engage the employees at that site. Members of the two committees may make up the membership of the sitewide safety committee.*

By using this sort of a committee structure, the efforts of the committees can be leveraged against each other for greater effectiveness. It also provides a meaningful conduit to senior management to the oversight committee or council. In some cases, this may be termed a policy committee as well.

Many large corporations have a health and safety committee made up of members of the Board of Directors as well to ensure that risk to employee safety are being identified and appropriately resolved. Also allows the Board of Directors to monitor the safety performance of the company.

Where multiple committees are needed, it is important to have a good structure in place to ensure the free flow of information both up this structure and down it. Such a structure also gives employees greater participation in decision-making that affects them when it comes to their personal safety or the risk to them and their coworkers.

## Safety Council/Policy Committee

In some jurisdictions, such high-level committees are required. However, it is good practice in a multi-committee environment to have a Council or committee to coordinate activities and solutions across the company.

Since this body is often a construct of the employer, it may not necessarily follow the same template when it comes to membership. The purpose of such councils is to monitor the activities of all the committees and share information across the committees. The Council also would act as an advocate for committees in resolving difficult issues.

As the name suggests, this body would have broad decision-making powers and be able to set company policy. The Council would also have significant resources at its disposal to focus on any issues they deemed of a serious nature. The membership of this body would generally be that of the senior management.

However, in a multi-committee environment, the health and safety council may include a member of each committee, normally one of the co-chairpersons. The Council would be more concerned with risks to employee's safety at a higher level and in a way that could affect the company broadly.

As with any committee, the safety council should have clear terms of reference spelling out those topics covered under the section on terms of reference.

## Board of Directors Safety Committee

Often a board of directors will also have a safety committee. This is often comprised of three members of the Board of Directors for the company. This committee often draws on information from a safety Council or policy committee, if such a body exists. It also draws information from the health and safety department.

Some general duties of the Board of Directors safety committee would be things such as:

1. Review any strategic initiatives for safety;
2. Set the long-term strategy for the company's safety system;
3. Review and monitor industry trends and company related safety trends;
4. Monitor legislative compliance;
5. Recommend appropriate training for board members are executives;
6. Review any serious incidents;
7. Review any regulatory visits or action within the corporation;
8. Review the outcomes of audits or any continuous improvement plans;
9. Review and make recommendations on annual performance targets; or

**10.** Monitor and make recommendations regarding the resources for managing safety within the corporation.

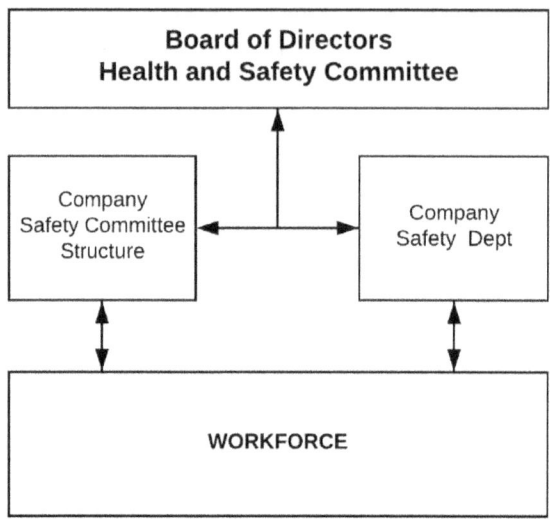

***Figure 6:*** *The committee structure represents another way that concerns can be brought forward throughout the management structure. The company's safety department and the safety committees have direct contact with the workforce and may be involved in addressing those concerns. Both sides would also provide information to the Board of Directors to assist in setting the strategy for the health and safety system and priority for resources.*

## COMMITTEE SIZE

When setting up a new committee, there may be a great deal of interest and many volunteers for the committee. Is often tempting to be as inclusive as possible, but large committees are unwieldy.

The size of the committee should reflect the size of the workplace. The smallest committee would consist of four members, two from management, and two from the employee side.

The size of the committee should be limited to a maximum of ten or twelve persons. Committees that are larger than this tend to become very inefficient. Larger groups have difficulty discussing issues and including everyone in a meaningful way. Additionally, large meetings are difficult to chair.

The greatest argument against large committees is that the meetings tend to take a very long time. It is important to recognize everyone's time has value, and wasting that time results in disengagement.

In large workplaces, a multi-committee system would work better than having one large committee. Limiting the size of the committee helps ensure the committee will be effective.

While a maximum number of members may be set out in the terms of reference, it is not necessary to fully staff the committee as long as the minimum requirements are met. It is better to have fewer, more engaged members than a full committee with several members who are not engaged.

## RECRUITING MEMBERS

Finding members for the committee usually involves finding members for the employee, or labor, side of the committee. However, when initially setting up a committee, it is important that appropriate management representation is obtained.

It is important to recruit members early. You must be sure that you have all the members in place or the replacement members in place at least a month before they will have to attend any meetings.

### Management members

Is important that the appropriate management personnel be initially assigned to the committee. While departmental, or area, representation is important for the employee side of the committee, the management side must also reflect the company's operations. It is important that management, or senior management, from the operations side of the company be represented in the committee as this is where most of the high hazard work would take place.

Recruiting personnel from the management side of the committee would normally involve meeting with them to outline what a committee does and determine if they would be interested in participating in such an endeavor. At least one of the management representative should be a senior manager in a leadership position, with access to significant resources.

It is important to obtain agreement from management representatives before their appointment so that they may become contributing members of the committee. This also allows the proposed members to obtain any required training before their appointment to the committee.

When setting up a new committee, it's important that the right personnel be selected. Representation should be from all major aspects of operations and have some representation from the safety department in an ex officio capacity.

## Union members

When initially setting up a committee, the person assigned this task may approach members of the union to determine their interest level in participating in a committee. Much like the discussions had with management representatives, it is necessary to explain how the committee should function and what the committee's role would be within the company.

The president of the union local, or chief shop steward should be approached to determine how the union would provide members to the committee. The union may appoint members or may have them stand for election at a union meeting. It is often a good idea that permission is obtained from the union for someone to give a short presentation at the meeting on what a committee would do and what their role would be so that people understand what they may be volunteering for.

Some unions may have set methods for appointing, or electing, members of for the committee. Such provisions are often found in the union bylaws. In some cases, a union may simply allow the employer to canvas for volunteers for the committee. This would mean using the same methods as stated for recruiting employee members.

Shop stewards or other union officials should not be appointed to the committee. This is in recognition that the safety committee is not meant to address, or deal with, labor relations matters.

Having union representatives (other than normal union employees), on the committee may lead the committee into involvement in matters best addressed through the union and the labor relations practices in place at the company.

## Employee members

When setting up a committee, or replacing members on an existing committee, broad and clear communications are essential.

The use of posters has limited effectiveness in recruiting committee members. Other methods have proven successful but require much more work.

1. **Email announcements.** Announcements over employee email may be effective in some workplaces. It is important that when sending out such messages that some sort of visual cue is included. This may be a photograph of the committee at work, or even a stock picture of a group of people working together. The message should stress that this is an opportunity to make a difference in the workplace. Your message should include some key things;

   a. what the committee does;
   b. what the committee has accomplished;
   c. why the committee is recruiting members;
   d. what members are expected to do regarding time commitments.

2. **Flyer in pay statements.** Putting a flyer in pay statements can also be successful. However, the flyer should be a single page only and clearly state its purpose. Refer to point number one for information that should be included in the flyer.

3. **Social media.** Many companies have social media accounts, and these are followed by employees. An announcement that the company is looking to get volunteers for a safety committee may not apply to all followers on social media.

However, it will demonstrate to all those followers that the company is committed to safety and has a working safety system.

4.  **Name a replacement.** Current members of the committee who are leaving the committee after their term may indicate to the committee that there is someone else in their area, or discipline, who would be interested in joining the committee.

5.  **Combine efforts.** While all the methods named above would have some success, using them all together would be much more powerful. A broad campaign would bring for the best volunteers and yield the best results.

## MEMBER EXPECTATIONS

### Representing an area or group

While clear expectations are often laid out for members in the terms of reference, these are often general in nature.

Committee members are often elected or appointed to represent the interests, or import, of a specific group or area. It is important that committee members do not lose sight of this important factor.

The committee members are not there to simply represent their views or champion their projects. While they may do this, they must also ensure that they are accurately representing the concerns and issues regarding the health and safety of those they represent at the committee.

Being a good representative means recognizing those situations where employees may be providing some information with an expectation that the person does something because they are a member of the safety committee.

Additionally, it is important that members of the safety committee actually solicit input from those they represent on topics being considered by the safety committee.

### Members receiving concerns/complaints

Often members of the committee may be approached by other employees and told of safety issues that are of concern to those employees. It is important that members recognize this for what it is. There is a clear expectation that committee members bring forward employee concerns to the health and safety committee.

Bringing forward concerns often means investigating the concern to ensure that the member understands the nature of the concern and any expectations from employees on what action ought to be taken.

In some cases, the concern would be of an urgent nature, and the expectation for the committee member would be to refer that to the relevant manager or safety advisor for quick action.

A committee member who has identified a concern, or complaint, should let the chairperson know so it can be placed on the agenda for the next meeting. This will let other committee members know that this issue has been identified and the member bringing it forward should be prepared to speak on that issue at the meeting.

**Soliciting input for the committee**

Before a committee meeting, members should be speaking to people in the workplace to see if there is anything that they would like you discussed at the meeting. This is also an opportunity to get input on anything that the committee is currently working on or discussing.

Committee members should ensure that they are using opportunities where they speak to other employees to get their feelings and import on anything that the committee may currently have under consideration.

## MEETING THEME

The idea of a meeting theme is not a new one. It is often something that is missed in more formal meetings. For those planning meetings, or responsible for a committee, it is important to understand that you are essentially dealing with volunteers for the most part.

It is often a good idea to have a central theme of for and meeting. That helps keep the meeting focused. The theme is often something new that is introduced in the meeting. It may be a project that the committee will work on, or the theme may simply be a review of incidents from a certain time period, such as the previous year.

For those who plan meetings, the central theme of the meeting helps engage members and lets them know what to expect in the meeting. Regarding safety committees, the meeting theme may also be termed a "primary focus" for the meeting.

Often meetings are reviewing what has already taken place, and there may be little in the way of new issues for the committee to consider. By establishing a meeting theme, the person planning the meeting can also bring in additional information that the members may find informative.

Examples of meeting themes may include:

- Review of company performance;
- Review of committee accomplishments;
- Review, or monitoring, of a specific program or process in the high-risk category such as:
    - Fall protection;
    - Respiratory protection;
    - Confined space entry; or
    - Critical lifts.
- Review, or monitoring, of company testing programs such as:
    - Atmospheric testing;

- o Hearing testing;
- o Workplace noise level testing;
- o Lead level testing; or
- o Other contaminant testing or monitoring.
- Committee presentations to the workforce;
- Committee recommendations to the employer; or
- Replacement of committee members whose terms are expiring.

## MEETINGS

### Preparing for a Meeting

The preparation done for a meeting starts well before the committee meets. The person assigned to set up a committee would be responsible for setting up the first meeting. A first meeting can be crucial to obtaining engagement in the committee. Proper planning will ensure that the meeting is done with purpose and remains on time, so that committee members will see that their time is valued.

### Providing information

Whenever a meeting is held, its success often hinges on the information that is provided before the meeting. The person in charge of the meeting must ensure that members are equipped for success. This means that at least a week before the meeting, members are provided with the appropriate materials.

Members should be provided with an agenda for the meeting. Additionally, membership should be provided with copies of any new documents that would be discussed at the meeting. This might include inspections, incident investigation reports, or any programs are processes under review.

### First meeting

The first meeting of the committee would normally be chaired by a member of management, which may be a management person appointed to the committee, or the senior safety person within the company. This is because the committee would not yet have elected co-chairpersons.

Often, companies tried to accomplish too much in the first meeting. The goal of the first meeting should be to ensure that everyone understands the requirements of the committee and its members. It should also be to elect co-chairpersons and to review the committee's terms of reference.

The person chairing this meeting should be prepared to step aside at the next meeting for the elected co-chairpersons.

The first meeting normally would take one of two forms. It would simply be a focus meeting on getting the committee up and running in ensuring that the members understand what their actual role within the committee will be and how the committee should function. This meeting should take no more than an hour.

The second form that the meeting might take is a learning session and meeting. In this, there may be several hours of basic familiarize a nation with committees and their role. This may take the form of a workshop. The last portion would be the first meeting of the committee.

**Agenda**

The agenda is often sent out by the person chairing the meeting at least a week in advance of that meeting.

Sending out the agenda in advance ensures that all the members can see what is going to be discussed at the meeting, and it also gives members an opportunity to propose changes to the agenda.

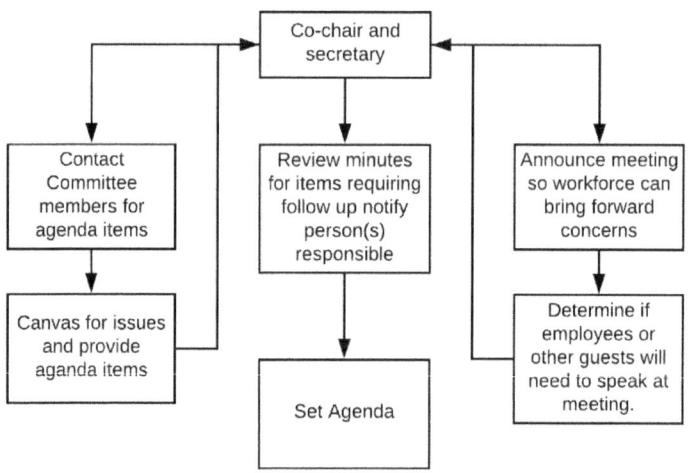

***Figure 7:*** *Setting the agenda for the meeting*

The agenda should have a specific format to help ensure that the meeting remains on topic and on time. The agenda should include several columns. As a minimum, the agenda would include columns for:

- Agenda item. The item to be discussed such as "quarterly incident review" and
- Sponsor. The person responsible for that agenda item. This is the person who will speak to that agenda item or lead the discussion on that agenda item.

The agenda should be descriptive enough to allow members to understand what things are being proposed for the meeting. Where a member of the committee is responsible for an agenda item, they should endeavor to work with the co-chair to ensure that background materials are provided to the members to save time in the meeting.

An agenda is very necessary as it forms the backbone of the meeting and gives focus to the meeting. It also allows the co-chairperson to gauge how long this will make the meeting. A heavy agenda may lead to an unproductive meeting, as the meeting may be too long.

The agenda should list all new business in descending order of priority. The highest priority item should be the first to be discussed. This ensures that the important items are dealt with at each meeting. If a meeting is running long, it may be necessary to defer agenda items. It is important to ensure that the high priority items are the first to be discussed.

The agenda should normally be only one page and effort should be made to provide hard copies of the agenda at the meeting.

One of the key rules about agendas is that if something does not appear on the agenda, it will not be discussed at that meeting. Provisions are normally made within the meeting to allow additions to the agenda, or amendments to the agenda, at the beginning of the meeting.

The agenda is an important tool to keep the meeting focused and on track. A meeting without an agenda is a meeting without a roadmap or clear purpose. It is this simple document that can promote effective meetings.

A sample agenda appears below.

### AGENDA

| # | ITEM | SPONSOR |
|---|------|---------|
| I | Review minutes of meeting of 12 Sept 1996 | Chair |
| II | Inspections | |
| | Inspection of substation #2 | Inspection team |
| | Outstanding inspection items | Joe Black |
| II | Review of accidents | Chair |
| | Review of accident action items | Chair |
| IV | New business | |
| | New safety boots | Joan Green |
| | Certification of boom operators | D. Hightower |
| | New metering guidelines | Meter supervisor |
| | Traffic control training | Fred Purple |
| V | Business from the floor | |
| VI | Adjournment. | |

Each agenda normally starts with a review of the minutes of the previous meeting. This is often termed "Old business." This is normally the first official item on the agenda and led by the chairperson of the meeting.

## Chairing a meeting

The two chairpersons usually take turns chairing meetings, and this is often scheduled well in advance.

The person chairing the meeting is normally responsible for setting the agenda, getting notification out to the members of the meeting and providing any background materials they may need at least a week before the meeting.

The person chairing the meeting should also attempt to communicate with the workforce to inform them about the things the committee will have under consideration for that meeting so that the committee representatives can solicit feedback from the workforce in general about those topics.

The chairperson should also review the meeting minutes from the previous meeting and be prepared to speak to each item on those minutes as they are to be reviewed at the next meeting.

Guest speakers, or employees who will speak at the meeting, also need to be informed about what to expect that the meeting and how much time they will be allocated to speak to the committee. Where possible, information on the presentations of guest speakers should be shared with the committee members.

## Length of meeting

The length of the meeting is of considerable importance in ensuring that the committee remains engaged in their tasks. Meeting should be limited, with whenever possible, to one hour or less. Ideally, the committee should meet for about an hour each time it meets.

It may be that the committee may need to have an unusually long meeting to accomplish certain tasks. In that case, a break should be clearly indicated on the agenda. It is often better to have two meetings of one hour each, then to have one long to our meeting.

Long meetings can lead to stagnation in the committee and the loss of effectiveness of engagement. While the one-hour rule is not a universal rule, each committee must consider the optimum time for a meeting. With experience, the chairperson can gauge how long a meeting may run, and so choose to defer some topics to the next meeting.

## Use of audiovisual aids

Committee meetings exist to involve representatives of the workforce in the monitoring and improving the health and safety system and the health and safety of the workplace. Meetings may benefit from the presentation of a video, or even a short PowerPoint presentation.

However, the purpose of the meeting is to discuss workplace health and safety issues or processes. Use of audiovisual aids is a good way to start a conversation. The overuse of audiovisual aids can stifle the activities of the committee.

Audiovisual aids are a good tool for presenting information quickly and clearly but should not be overused. The purpose of the use of such things is to bring about informed discussion.

## THE MEETING PROCESS

The actual meeting itself is the heart and soul of committee activities. It is important that meetings have a focus, which is normally provided by a strong agenda.

The primary responsibility of the chairperson at a meeting is to keep the meeting on track and avoid distractions. This would have meant that the chairperson would've distributed an agenda and any background materials that members would need to participate in the meeting.

Any employees, or other speakers/presenters, should have been spoken to by the chairperson to make clear the expectations around the amount of time they would have and what they would be speaking about.

Meeting should follow a structured format.

### Call to order

The meeting is called to order by the chairperson and the time of the meeting being called to order should be noted in the minutes.

In calling the meeting to order, the chairperson must assure themselves that there is a quorum and the meeting can proceed.

Most meetings are informal affairs but usually follow the general rules of the meetings such as those found in "Roberts rules of order." These rules are for very formal meetings but really provide a good structure for any meeting.

The most important aspect of these rules is that members make "motions." These motions are suggestions to change the agenda or to suggest that the committee undertake a certain course of action. It is up to the chairperson to see who support such a motion through another member seconding the motion and looking for consensus among the members.

The committee may not use the term "motion," but the concept would still be applicable in that members may suggest things and it is up to the committee chairperson to gauge consensus. This is often done by asking each member of the committee the state their thoughts on what is being proposed.

The formality of the meeting is normally dictated by the corporate culture of the company, rather than any formally established rules. It is important to reflect that the meeting is of an important nature and that the agenda is the roadmap for the meeting.

## Determine attendance

The first part of any meeting is to determine who is in attendance and record that in the minutes. Committee members should communicate directly to the chairperson if they are unable to attend a meeting so that meetings are not held when the establishment of a quorum is unlikely.

Those not in attendance, who have communicated their inability to attend should be recorded as not in attendance at the meeting with "regrets." Members not in attendance who have not communicated their inability to attend should be contacted by the chairperson to determine the reason for their non-attendance. The reason for this is that it helps the chairpersons determine the engagement level of committee members.

## Review the agenda

It is important for the chairperson to review the agenda at each meeting. This may not entail reading through the entire agenda to those in attendance but verifying that those in attendance have reviewed the agenda.

The chairperson should ask if anyone would like to add or amend anything on the agenda. This is normally the last opportunity for members to insert something into the agenda for the meeting.

The reason for this is so that it is clearly understood what will be discussed in the meeting and will allow the chairperson to ensure the meeting remains on track and does not stray off in the topics that are not on the agenda.

There may be some time between the drafting of the agenda and the meeting, so it is important to allow members to add to the agenda before the meeting really kicks off. It also gives an opportunity to amend, or withdraw, agenda items if the information has changed or things have not met their schedule.

## Proceed with meeting

In proceeding with the meeting, the chairperson would start off with the first agenda item, which is normally "Old business." Old business is simply a term for items already brought before the committee and not yet resolved.

As with the agenda, each item in old business will have a person responsible for it, or a sponsor. Reviewing old business gives the people responsible for these items a chance to update their status. Items brought before the committee should remain in the minutes until they are indicated/annotated as being completed, or resolved, within the minutes.

At each meeting, the outstanding items before the committee should be reviewed to ensure that they do not become forgotten. It is often up to the chairperson to make sure they have a clear understanding of why items that are long-standing, but have not been completed, or resolved.

It is important to understand that the committee has a role of assisting and advising the employer on safety matters. The committee also has an important role in the workplace of representing the concerns of employees. Items may come before the committee that may not be implemented, or successfully completed.

In some cases, the resources may not be available, or the employer may not support the action proposed by the committee. These items may be marked as "resolved." Items that are resolved should contain the rationale, or brief explanation, for not completing the action.

In circumstances where the employer declines to undertake actions or activities that the committee feels are required, the committee may decide to make a written recommendation to the employer explaining their rationale for the recommendation and the benefits the recommendation would bring. This will be covered in a later section of the book.

**New business**

Most meetings are divided into two distinct parts; old business and new business. New business is things that the committee is discussing for the first time. Normally most of the meeting is spent on new business.

Old business may include standing agenda items such as the review of inspections or incidents, or these may be included in new business. Old business is always dealt with first so that the committee may know what matters are still under consideration and what matters are still unresolved.

New business would be things like guest presentations, employees coming to the meeting to express concerns, or new projects or task being undertaken by the committee. The committee may, from time to time, undertake projects to review certain aspects of the safety management systems such as a training program or monitoring program.

The committee may also conduct surveys or review current company action plans regarding health and safety that have resulted from audits or incident investigations.

## Presenting agenda items

The chairperson is responsible for moving the meeting forward and establishing consensus in the committee. For each agenda item, the chairperson would yield the floor to the person responsible for that agenda item. During this time, it is also the responsibility of the chairperson to ensure that the meeting remains on topic.

Agenda items may give rise to the need for the committee to make a decision or to establish a consensus. The chairperson would normally do this by polling the members of the committee.

Polling simply means asking each member individually to give their viewpoint on a specific topic. In this way, all those members of the committee are heard. It is simpler for the chairperson to recognize where the consensus, or majority agreement, lies.

## Staying on time

It is often not possible to determine exactly how much time will be spent on any single topic. A properly organized agenda will contain the topics in descending order of priority. The reason for this is that the chairperson may ask that the meeting be brought to a close and some agenda items deferred until the next meeting.

Deferring agenda items to the next meeting should not be a regular practice but may be necessary in some cases where discussions or presentations go on too long.

It is important to stay on time to demonstrate to members that their time and their contribution is valued in the committee.

## Bringing the meeting to a close

The chairperson may ask for a motion, or simply bring the meeting to a close when all the agenda items have addressed.

In the case of the meeting running very long, the chairperson may suggest, or make a motion, that it the meeting and. The chairperson should not arbitrarily and the meeting. Making a motion allows members to agree that the meeting should end and ensures that the meeting is ended only after committee members feel that they have properly addressed the current agenda item.

## Setting the next meeting

The final agenda item in any meeting should be to verify the date and time of the next meeting. Committees often meet on a set day and time, but it is important to verify this as some members may indicate they are unable to attend at that time.

If the date and time of the next meeting are reviewed before the closure of the meeting, it will be possible for the chairperson to propose a new date and time if many members are otherwise engaged.

## MEETING MINUTES

The meeting minutes are a record of what the committee discussed or during the meeting and what decisions, or resolutions, were made by the committee.

The minutes are also an instrument of communication with the workforce, as they should be available to all employees for review.

### Secretary

The secretary is the person who normally would create and update the minutes for the committee. The Secretary may be an administrative person provided by management to provide support to the committee.

In smaller committees, the co-chairperson that is not chairing the meeting may act as the secretary. Where dictated by necessity, the person chairing the meeting may sometimes act as of the secretary.

It often falls to the secretary to keep the minutes and to distribute the minutes as required. Once the minutes are completed, they are usually circulated to both co-chairs for review and signature. This is to show that both co-chairs are involved in committee business and that both have approved the minutes as an accurate representation of the operations of the committee and its stance on the various issues it is considered.

### Formatting the Minutes

To be effective, the minutes of the committee must contain certain elements. The minutes should contain:

- The date and time of the meeting;
- Who was, or was not, in attendance (including guests or other presenters);
- The name of the chairperson;

- The items discussed;
- The status of the items discussed;
- The person responsible for the items discussed;
- Target dates for completion of any items discussed;
- The date of the next scheduled meeting; and
- The signatures of both co-chairpersons

A sample format for committee minutes appears in the appendices.

The formatting of the meeting minutes is explained in detail in the appendices.

## Distributing the minutes

Normally the secretary would be responsible for the distribution of the minutes. Distribution of the minutes can be as simple as posting them on an internal website are making them available on the server. The distribution of minutes should be addressed in the terms of reference.

In many companies, the minutes are posted on a bulletin board with other safety documents or notices so that employees might review them when they wish. The company must make the minutes available to employees in a convenient way. Normally minutes are made available electronically and in hard copy.

Some companies may post only a few months worth of minutes. In posting one set of minutes, an older set is removed so that there is not a huge accumulation of safety committee meeting minutes on the bulletin board.

Any presentations or background material that was given to the committee should be attached to the minutes and made available to those who are reading the minutes. Exceptions are personal information, detailed incident investigation information, or other information that should be kept private.

## COMMUNICATING WITH THE WORKFORCE

Many committees make the mistake of simply sending out their minutes and fail to communicate with the workforce properly.

Most committees have a significant influence on the resolution of safety issues and are involved in the review and development of the health and safety system. Unfortunately, many employees do not understand the committee's role or what the committee accomplishes.

The purpose of the meeting minutes from the health and safety committee is so that the committee itself can keep track of issues that are brought before it and how those issues are resolved.

Unfortunately, to meet those requirements, the minutes must be structured and so do not make for a very interesting read for those who do not have a real stake in the issues before the committee.

### Issuing a Synopsis

A synopsis is simply a short summary that is brief and easy to read but still communicates meaningful information. Instead of a multipage document, like the meeting minutes, a synopsis can be a single page giving highlights of the committee's activities.

A committee that meets every month may choose to issue a quarterly or a semiannual synopsis of his activities. This can include some issues of interest to the employees that are still under consideration. It can also include some numbers, such as the number of issues that came before the committee during that period and year to date.

A synopsis can show the workforce that the committee is having a positive effect on the health and safety of the workplace and is an important partner in improving workplace safety. It can also show the committee as a credible alternative when traditional methods of reporting concerns or hazards are not being answered to employee satisfaction.

The committee can issue a synopsis, an update, or even a stewardship report to show that it is working towards a safer workplace. The basic communication can include things like:

- inspections completed;
- issues raised;
- incidents reviewed;
- issues resolved; and
- any measures that the committee might use to demonstrate positive performance.

Releasing a synopsis would mean more than simply painting into a bulletin board. This tool is most effective when delivered by a senior manager and an employee gathering or meeting. The committee may also use email or social media to promote its work.

In the case of a synopsis, **it may be necessary to get approval for the release** of any information in the synopsis that may be seen as sensitive or identifying specific people or departments.


**Verbal Briefings**

One of the most common methods that a committee promotes its activities is through verbal briefings, and other meetings are employee gatherings by a member of the committee.

In larger organizations, verbal proof briefings can be used to communicate the update or synopsis made by the committee to ensure consistent messaging. In smaller companies, a verbal briefing is a good way for a committee member, or the co-chair, to communicate to groups within the workforce what the committee is working on and what the committee has accomplished.

## Intranet and Social Media

It is common in larger companies for safety committees to have a webpage on the internal network. Normally the safety committees within an organization share a single webpage.

Using the internal network makes the distribution and retention of minutes much easier. It also ensures that minutes and associated material are readily available for reference by any employee.

Social media can be used to highlight the committee's accomplishments. Where a company has its own Facebook page and even a Twitter or Instagram account, this can be used to highlight the committee's accomplishments.

Within the framework of corporate social responsibility, the company can demonstrate that it is listening to employees and resolving issues that arise in the workplace. Using social media would only be appropriate where the committee has met a significant milestone or accomplishment. It would not be the appropriate venue for a quarterly update or synopsis.

## Annual Report

The committee itself should review its performance annually to determine areas where the committee has been effective and successful. It should also attempt to identify areas where the committee should gain more insight, expertise, or become more active. In some cases, it may be necessary to devote an entire meeting to review the year of activities and what the committee can do to improve.

This represents an excellent opportunity to engage the workforce as personnel can be invited to attend the meeting to give feedback to the committee and to make suggestions for improving the committee from their perspective.

The output of the annual report can be a summary of the committee's accomplishments over the past year, and the strategic plan for the committee for the upcoming year. This would normally be compiled by the co-chairpersons and reviewed by the committee before release.

## INVOLVEMENT IN SAFETY ACTIVITIES

Most committees are mandated, by legislation, to be involved in certain activities in the workplace. The most common things that committees would be involved in our:

- Workplace inspections;
- Incident investigations;
- Program review;
- Review of testing, or monitoring; and
- Work refusals.

### Workplace inspections

The committee's terms of reference should outline how the committee will be involved the workplace inspections. There are several different options on how the committee can be involved.

It is important that the company's inspection process reflects how the safety committee would be involved in inspections within the company. This could define the method used by the committee to exercise its inspection responsibilities and how the committee would receive information relating to inspections in the workplace and any deficiencies that had been identified for resolution.

### Committee as the inspector

In some cases, the committee will conduct inspections immediately before or after a meeting. Members of the committee may be selected or designated to conduct these inspections.

It is very appropriate that the committee would provide copies of these inspections to whoever would oversee gathering that information and ensuring follow-up. This would mean that the committee would have to use the company-specific inspection process to do this.

For inspections conducted by the committee, it is important that the deficiencies be tracked in the committee's minutes until they are resolved.

## Committee participating in inspections

The committee may choose to participate in workplace inspections in another fashion by taking part in regularly scheduled inspections. The committee may assign one or more members to accompany personnel on their regularly scheduled inspection tours.

Where a committee member accompanies an inspection, this should be noted in the committee minutes, and a copy of the inspection report should be provided as an appendix to the minutes.

## Committee reviewing inspections

In some cases where a committee represents a smaller workplace, the committee may choose simply to review the monthly inspections. This would be impractical in larger workplaces as would be many inspections for the committee to review.

Reviewing the inspections allows the committee to determine if the inspections are being done diligently and for the committee to make recommendations on findings or to further investigate findings of the inspection.

## Committee and incident investigations

The involvement in incident investigations should be addressed in the committee's terms of reference, and in the incident reporting and investigation process in the safety management system.

Virtually all committees receive a summary of incidents provided by safety personnel. The committee will also likely receive a summary of near miss or hazard reports if the company has such a system.

## Committee as an investigator

It is a common practice to involve a member of the committee in the investigation of serious incidents, or incidents that are designated as high risk. The committee should clearly understand who will be involved in such investigations. Often it is one of the co-chairs that would participate in incident investigations.

It would be up to the committee to ensure that any member participating in incident investigations had some specific training in the incident investigation process and the cause methodology utilized by the company.

The inclusion of committee members should be reflected in the company's incident investigation reporting process to ensure that the appropriate members of the committee are notified when there has been a serious incident so that they can participate in the investigation.

The representative from the committee would be part of an investigation team and not the lead investigator. Since the committee exists in the advisory and oversight capacity, it may not be appropriate for a member of the safety committee to lead an investigation on behalf of the company. In some cases, safety committee members may be empowered to undertake their own investigation or lead an investigation.

## Committee as a reviewer of investigations

Instead of being directly involved in incident investigations, the committee may choose to simply review investigation reports provided by the company. The reason for committee members not participating directly in investigations can be many, and each committee must determine what would be appropriate in their workplace.

Some workplaces in the transportation industry may have an incident review committee for transportation incidents that would be responsible for reviewing transportation incidents and so make the committee's involvement in the review process incidental.

The committee may review incidents to ensure that the report is clearly written, and the incident can be clearly understood by the reader. Additionally, the committee would be reviewing the incident to ensure that appropriate causes are identified and that the recommended action is appropriate to address those causes.

The committee may relate identified causes, and corrective actions in incident investigation reports. They may also identify outstanding or emergent issues within the workplace and choose to make their own recommendation to management.

One of the important aspects of the committee as a reviewer of incident investigation reports is to ensure that the recommendations are resolved and closed out. The committee may also target inspections to ensure that these corrective, or preventive, actions have been effective in eliminating the identified underlying, or root, cause.

In its role as a reviewer of incident investigations, the committee may ask you the company for additional information surrounding incidents. The committee may also ask questions regarding incidents to ensure the investigation did address all aspects of the incident and that the follow-up is being done to ensure that the identified causes are being mitigated.

## Program review

One of the key benefits for the employer in having a safety committee is that they may be a sounding board for new initiatives or improving existing processes and programs.

Committees may review programs from time to time to ensure that they are effective. This is often done using the written process and information provided by the employer. In some cases, the committee may undertake a survey of the workforce to determine the relevance and effectiveness of programs or processes.

The employer may come to the committee and asked the committee to review process changes as part of a Management of Change (MOC) process were simply to ensure the committee has an opportunity to review any proposed changes and make suggestions for improvement.

In some cases, the program or process that is being reviewed or contemplated is complex. The committee may form a subcommittee, with a subject matter expert, to review the programmer process and make a report to the committee.

## Review of testing, or monitoring

Committees often are empowered, through legislation, to review plans for testing within the workplace and to review monitoring programs that are already in place.

The purpose of this is to ensure that the committee understands why testing is being conducted and how it is meant to be conducted. This would allow the committee to communicate with its members and the workforce about the method, and purpose, of any testing.

It is expected in such cases that the committee would receive the results of any testing for review and recommendation. Testing is often conducted in the workplace for chemical contaminants or hazardous noise.

Some workplaces have a monitoring program to ensure employee safety. Monitoring programs may take the form of dosimeters of for radiation, or blood tests for lead contamination. The committee is usually provided with the monitoring reports for these company monitoring activities to ensure that the monitoring is being done and that no noteworthy results have been missed.

The committee may, from time to time, request monitoring from the employer to determine the qualities of a hazard present in the workplace. The most common example is noise monitoring.

## Work refusals

Work refusals are quite rare. In such cases, emotions can run high, as a work refusal is normally the result of the employer's inability to properly address safety concerns with an employee.

The committee's involvement in work refusals a should be outlined in the committee's terms of reference, and in the company's work refusal procedures. Most companies properly classify a work refusal as a serious incident and require an investigation.

The purpose of the investigation may not be just to invalidate, or validate, the concern raised by the employee. The investigation should also address why the work refusal became necessary from the employee point of view.

The involvement of committee members and work refusals is often described as a requirement in legislation. However, in any case, a committee member from the employee side of the committee should be involved in the resolution and investigation of any work refusal.

This will help establish in the minds of employees that they will be treated fairly in such cases and that the company takes such cases very seriously. The involvement of the committee will also bring an objective third party to the situation and prevented from escalating unnecessarily.

## Harassment and Bullying

In some jurisdictions, harassment and bullying are in the realm of health and safety as workplace hazards.

The committee has a role in ensuring that the company has an appropriate harassment reporting and investigation process. However, this is not something that the committee would normally be directly involved in.

The reporting and investigation of harassment in the workplace may, or may not, be seen as a safety responsibility. In some companies, this responsibility rests with the human resources department, but that does not mean it is outside the committee's purview.

While the committee would not usually be involved in specific incidents reported by employees, or the investigation of those incidents, the committee may receive reports on the number of reported cases of harassment and the results of investigations.

Out of respect for those involved, the information would have to be closely monitored to ensure that identifiable information is not included in any reports provided to the committee.

The committee would also have to be careful in communicating any statistics to the workforce regarding this sensitive area, and should seek guidance from the employer before any such considerations are made.

It would be appropriate for the committee to make recommendations to the employer regarding the handling of harassment complaints and investigations or in responding to any surveys that identified a high number of incidents of harassment or bullying in the workplace.

In cases where regular employee engagement surveys show rising incidents of harassment or unwelcome behavior, it would be appropriate for the committee to recommend some action be taken to verify these findings and to address specific issues in order to curb the problem.

In serious cases of reports regarding harassment or bullying, it is often prudent for the company to engage a third-party investigator to ensure that the matter is dealt with objectively. Most companies are not well-equipped to deal with more than a few complaints per year.

The role of the committee in this area would be less direct than in other areas. However, it is appropriate for the committee to ensure that there is a system in place that is working and that is functional.

**Harassment or bullying complaints**

Since harassment and bullying may be in the realm of workplace health and safety as a workplace hazard, depending on the legal jurisdiction where the committee operates, the committee may receive direct complaints of harassment or bullying. It is important that the committee define how they will handle such complaints within their terms of reference.

The committee cannot refuse to accept harassment or bullying complaints. However, it is problematic for the committee to validate such concerns and complaints. The committee must keep any reports of harassment or bullying confidential.

The company harassment and bullying reporting and investigation process should address how complaints to the committee will be treated and how the process will respond in terms of referring complaints directly into the appropriate reporting system for resolution.

It is appropriate for the committee to track the number of such complaints that it might receive, and it is also appropriate for the committee to follow up with the employer to ensure that these complaints have a had been investigated to establish what further action would be required.

It is not advisable that any member of the safety committee participate in a harassment or bullying investigation unless there is a work refusal involved.


**Workplace violence**

Workplace violence is an area of growing concern in our society. The committee would certainly be involved in reviewing the company's process of assessing the risk of workplace violence and the measures taken to prevent workplace violence and to respond to incidents of workplace violence.

The basic expectations around workplace violence are that the employer would conduct a risk assessment, in consultation with the committee. This is the standard practice in most areas of the safety management system.

The committee in its capacity as a monitoring and assistance body can validate the process, or suggest changes. The committee can also be a catalyst for ensuring that workplace violence drills are held where that is appropriate.

It should be noted that workplace violence can have both internal and external sources. Workplace violence is not bullying but involves some threat of injury or harm.

This can mean a threat of physical harm or even psychological harm.

Incidents of workplace violence are not very common. However, the committee should treat information around workplace violence in a similar nature to any other personal information it receives in that the information should remain confidential and no information should be communicated that would readily identify individuals involved.

## COMMON PITFALLS

There are common reasons why committees are not successful and become ineffective and boring. Some of these are much more obvious than others.

Below is not an all-inclusive list, just some of the common pitfalls that befall well-meaning committees.

### Safety persons are chairing, or on, the committee.

It often seems intuitive that safety people need to be involved with the health and safety committee. That is certainly true in that they need to be involved with the committee.

However, the committee is there to assist the employer and act as a safeguard separate from the safety function within the company.

Because of their greater knowledge and experience in the world of health and safety, the safety personnel may inadvertently overshadow the committee and ultimately direct the actions of the committee.

The opposite is the desired outcome. The safety committee should be providing some and direction, and requesting services from, the safety function.

### No clear terms of reference

Committees are often cast adrift once they are formed. Without a clear terms of reference, the committee has trouble communicating to others what its purpose is, and even understanding that purpose themselves.

A clear terms of reference is absolutely essential to defining how the committee will operate. Often when a committee becomes stale, it is because they have let their focus shift away from their primary functions of representing employees, reviewing management's efforts to provide a safe workplace, and participating in key functions like incident investigations.

## Poor communication with the workforce

This is a common issue where the committee relies simply on minutes, posted or distributed, to inform the workforce of its efforts.

The committee must do more than simply distribute minutes. The committee must communicate important issues to employees using bulletins, and other mean suggested earlier in this book.

The committee must also give meeting highlights at appropriate safety meetings with employees to demonstrate that the committee does resolve issues and does monitor the employer's health and safety program or system.

Like any group, the committee must celebrate and share successes.

## Poorly defined member expectations

It is incumbent upon the chairperson to ensure that new members understand their obligations to the committee. In some cases, members may not participate in the committee to the appropriate level because they do not understand the requirements.

This is often related to having a poor terms of reference that does not clearly spell out the expectations for committee members.

It may be necessary in some cases to have new committee members review their responsibilities and sign an agreement to meet those responsibilities.

In extreme cases, the committee may recommend that a member is removed from the committee so that a more active member can be appointed or elected.

## Committee is overwhelmed

New committees can often be overwhelmed by the perceived requirements to participate in many areas of the health and safety system.

In these cases, it is incumbent upon the chairpersons to limit the committee's activities to what can reasonably be accomplished.

The committee should take a long-term view of its role. This does not mean reviewing incidents at every meeting, or inspections for that matter. The committee may designate certain meetings that have a specific purpose and only deal with other issues of an urgent nature at those meetings.

## Committee members are not sufficiently trained

To be effective, the committee does need some expertise. Sometimes committees are set up before even the Co-chairpersons can be adequately trained. Training should be a priority for committees, particularly for Co-chairpersons.

Setting up committees requires support. This includes resources to provide basic training to Co-chairpersons. Co-chairpersons play a significant role in the effectiveness and sustainability of the committee. Recommended training appears earlier in this book.

## Committee becomes involved in areas outside the mandate

To be effective, the committee needs to remain focused on its mandate.

Unfortunately, there are times when people may come to the committee with issues more properly directed within the labor relations or human resources realm. They normally do this hoping for a more rapid resolution to their concern, but this only serves to bog down the committee.

It is important that Co-chairpersons guard against this. This is one of the reasons that the agenda should be vetted and approved before being approved as it is important to understand what any visitor to the committee wishes to discuss.

Often this is caused when persons from the union that have duties relating to the union, such as shop stewards are appointed to the committee or elected to the committee. It may be that the union has not recognized the potential conflict of interest that exists in this case.

The committee does have the purpose of assisting the workforce in establishing a safer workplace, but the committee has no business becoming involved in labor relations issues such as grievances or discussing employee entitlements related to human resources policies.

These issues should be directed to the appropriate departments and agencies within the company and studiously avoided by the committee.

Involving the committee in these areas would result in the committee wasting its reason horses, and the time of its members, in areas where it really has no authority.

In some cases, amending the union practices around appointment or election of committee members may be necessary to ensure that those appointed, or elected, to the committee do not have other duties that may draw their interest away from the committee's focus.

**No mechanism for worker input**

All committees need a clear mechanism for worker input. This may include active solicitation of input, surveys, or other means.

Committees sometimes rely too much on the hope workers will come forward with issues or concerns. This is often not the case. Waiting for input may lead to a stagnation of the committee.

Committee members must *actively* solicit for input and investigate concerns *before* a meeting. The members of the committee should accept verbal reports of concerns or issues and investigate them to determine the facts where possible.

Committees may also use surveys to get input from workers. Surveys are often not favored with a high response rate, and so they should not be one too often. Some companies actually do workforce surveys, and it may be possible to become part of that process.

If surveys are used, then some potential reward, such as a prize draw, is effective at boosting participation. Remember that such input does have value and can raise the profile of the committee.

The danger with surveys is that there is an expectation that the results be reported back to the workers and that something is done. Adequate resources should be on hand to handle the reporting back to the workforce and investigating, then prioritizing the issues identified.

Once the results of a survey are communicated, a plan should follow. The committee (and the employer) should show that the input is valuable, and the legitimate concerns are being addressed in a priority fashion.

This also provides the opportunity for more positive communication with the workforce as the committee reports on progress as the plan unfolds and concerns or issues are addressed and resolved.

It is often important to explain the rationale for not acting or not being able to resolve specific concerns or issues. In some cases, proposed actions may be unworkable or simply too costly. This still requires an explanation for the workforce.

## REVITALIZING A COMMITTEE

Like anything else, committees can move back and forth between effective and ineffective.

The reasons that a committee becomes ineffective or stale are varied. The following is meant to give some strategies for revitalizing committees to re-engage members or to re-engage the workplace.

### Members have been on the committee for too long

People volunteer to be on committees for a variety of reasons. In some cases, once their particular motivation for joining the committee has been resolved, members may lose interest in the committee but feel obligated to remain.

In some cases, members may feel being on the committee gives them some prestige in the workplace, and so they cling to committee membership even though they no longer feel motivated by the committee's activities.

In looking at the membership of the committee, it is important to look at members who have been on the committee for more than one term to see if they still are contributing to the committee and understand their reasons for remaining on the committee.

Often, new members can breathe new life into the committee with their different viewpoints and different priorities.

### The committee has difficulty reaching consensus

A committee works by consensus and getting consensus in a larger group is often much more difficult than in a smaller group. It may be that your committee is just simply too large. Your recruiting efforts have been too successful, and there are too many people on the committee.

Most committees only meet monthly, or perhaps less frequently. With the frequency of meetings and their duration, it is often very difficult to gain consensus on key issues with a group of 12 or more people.

This is one of the reasons that the terms of reference should define the size of the committee. If the committee seems to be necessarily large, it may be advisable to split the committee into two different committees.

Smaller groups engage with each other much better and are more easily able to find consensus.

## The committee lacks direction

Every committee may suffer from this problem from time to time. Even successful committees would suffer from a lack of direction after the completion of a major project.

Some committees tried to have an annual goal or target. They may even have more than one goal or target to help give the committee direction. This does help keep the committee focus.

A good terms of reference only specifies how the committee would function and really does not give the committee specific targets. However, where a committee lacks direction, it should look to the functions it is meant to fulfill.

A committee that needs a focus may be guided onto a project to review an employer process in a high-risk area. The committee may also undertake engagement activities such as a survey of employees to determine where its priority should lie.

There are many opportunities for activities and projects to refocus the safety committee. Conducting a quality review of incident investigation reports done within a certain period of time we engage all members of the committee and provides meaningful feedback on the incident investigation reporting process itself.

Committees that lack focus and direction often need a project. The specific project, or goal, is most beneficial if it comes from the part of the workforce that the committee is meant to represent.

Lacking any specific direction from the workforce, the committee should return to its primary functions and focus on one of those areas in looking at things like:

- Testing or sampling in the workplace;
- The effectiveness of workplace inspections;
- The effectiveness of company training;
- Incident investigation reporting;
- Workplace orientation;
- Emergency response planning and drills;
- Mechanisms for workforce input into committee activities;
- Review of risk assessments; and
- Validation of critical task lists. In many cases, employers have a list of critical tasks. Critical tasks are those tasks that must be done with a high degree of accuracy each time in order to avoid significant negative consequences. These are usually identified because of their high risk.

**People are reluctant to volunteer**

This issue often arises where committees are viewed as not being particularly effective in the workplace.

You should refer to the section in the book on recruiting members. However, people failing to volunteer may indicate that the workforce perceives the committee as ineffective or does not clearly understand what the committee does.

To interest people in volunteering for the committee, it must be seen as an effective body, and the workforce must clearly understand the purpose of the committee. If there is a sense that people are reluctant to volunteer for further research must be done to determine the general reason for this reluctance.

With an established committee, when looking for volunteers, it is important to lay out what it is the committee expects to be working on in the next year and what the committee does hope to accomplish.

This is a good way to pique the interest of those who may be interested in volunteering but may be reluctant because they don't understand what the committee is actually doing.

When faced with a general reluctance to engage with the committee, it is recommended that a survey be conducted to determine employee perception of the committee. It may be that the committee is suffering from an image problem. One of the most often neglected areas of safety committees is the committee's communication with the workforce.

A call for volunteers should likely include:

- Why there is a position open;
- The committee's accomplishments over the past several years;
- The length of the term to be served;
- The basic expectations regarding time and effort; and
- The focus or targets of the committee for the coming year.

### Committee members feel trapped on the committee

This is not an uncommon occurrence. This is often referred to as the "submarine service" option. Long ago members of submarine crews could only leave the vessel if they found someone to take their place.

This reason feeds back to the first problem, which is members have been on the committee for too long. If members feel they cannot leave the committee, they are unlikely to be motivated.

In these cases, it is necessary for management and the safety function to show, and demonstrate, their appreciation for the members of the committee and validate the importance of their work.

The company must work with its employees to make the committee a desirable thing to be involved in the workplace. This is not always a successful effort.

In cases where members really do feel that trapped being on the committee, there is another problem to solve. The company may be required to have such a committee, and in small companies, it is difficult to rotate members in an out of the committee.

In a larger company, a concerted effort by management, the committee, and safety function would normally be successful in recruiting additional members. However, in smaller companies, this may not be successful for a wide variety of reasons.

Best practice is that employees volunteer for the committee where possible. The rationale behind getting volunteers is obvious. Volunteers actually want to be on the committee. In extreme cases, where there is no union, the employer may choose to appoint employees to the committee.

Appointing employees to the committee is not the preferred option, but when no other option is available, this is occasionally done. In such cases, consideration should be given to reducing the length of the term someone would serve on the committee and clearly communicating that this was the only option left to the company.

If the company chooses to appoint members to a committee, it is extremely important that the committee has a specific focus, projects, and targets. Even appointed members may realize that the committee is engaged in important work and that it is worthy of their time.

**The committee seems disconnected from the workforce**

Committees may have their own priorities. However, the perception may be that the committee is disconnected from the employees is meant to represent and that it is not responsive to concerns brought forward.

It may also be that the committee is receiving very few concerns or reports of issues even though incidents continue to happen in the workplace.

Even effective committees may have the perception that they are somehow separate from the workforce and a mysterious sort of entity. This can happen unintentionally, but it can be difficult to address.

One of the best ways to socialize the committee is to take advantage of some of the many events throughout the year that mark safety in the workplace or recognize labor. These weeks recognizing workplace safety or holidays recognizing the workforce are good opportunities for the safety committee to sponsor and hosts a social event with the assistance of management.

These events should be appropriate for the workplace and the workforce. These can be as simple as a barbecue, or even a small trade fair.

Important to the success of any attempt to socialize the committee is that there be as few speeches as possible and as few expectations on attendees as possible. It would give the members of the safety committee an opportunity to interact with members of the workforce in a relaxed setting and explain what it is the committee is doing and what his current focus is.

Depending on the workplace, it may also be prudent for the members of the committee to be identified in some way using a shirt that identifies them as a member of the safety committee or some other means.

In order to really connect with the workforce, the committee must understand and communicate that it exists to serve the workforce and to enhance the employer's safety system (clear purpose). It does this through monitoring that system and also by giving a voice to the workforce to express concerns and identify emergent risks and hazards in the workplace.

## The committee engages in business not appropriate to the committee

While a safety committee is meant to deal exclusively with matters related to the health and safety of employees, this is a broad mandate.

It is often the case that the committee and allows itself to be drawn into things better handled within the realm of human resources or labor relations.

Employees may bring complaints or concerns to the committee that relate to collective bargaining issues or other employee entitlements.

It often falls to the Co-chairpersons to ensure that the committee remains focused on its mandate rather than other issues. Where this is an identified problem, and the committee has strayed from its mandate is often necessary to hold a special meeting. The special meeting would be the committee members only and would seek to identify these occasions when the committee had strayed from its mandate.

It is important to act quickly in these cases as these other issues bog down the committee and make it much less effective. Ultimately the committee has no authority in those areas and often people seeking to circumvent existing systems will come to the committee hoping for some quick action.

**Appendix 1 Simplified Sample Terms of Reference**


NAME OF COMMITTEE

1.  The committee shall be known as the Green Safety Committee

AREAS REPRESENTED

2.  The committee shall represent all employees working in the Transportation hub facility.

FUNCTIONS OF COMMITTEE

3.  The functions of the committee are:

    (a) to identify, evaluate and recommend a resolution of matters pertaining to health and safety in the workplace to appropriate managers;
    (b) to encourage adequate education and training programs in order to ensure that all employees are knowledgeable in their rights, restrictions, responsibilities;
    (c) to assist and consult in the development and maintenance of the Health and Safety program of the organization;
    (d) to promote Health and Safety in the workplace;
    (e) to make recommendations to the employer and the workers for the improvement of the health and safety of workers;
    (f) to recommend to the employer and the workers the establishment, maintenance and monitoring of programs, measures and procedures respecting the health or safety of workers;
    (g) to obtain information from the employer respecting the identification of potential or existing hazards of materials, processes or equipment,

(h) to obtain information from the employer concerning the conducting or taking of tests of any equipment, machine, device, article, thing, material or biological, chemical or physical agent in or about a workplace for the purpose of occupational health and safety;

(i) to be consulted about, and have a designated member representing workers be present at the beginning of, testing conducted in or about the workplace if the designated member believes his or her presence is required to ensure that valid testing procedures are used or to ensure that the test results are valid; and

(j) to participate, through designated certified members, in the work refusal or stoppage process.

## COMPOSITION OF THE COMMITTEE

4.     The committee shall have management and employee Co-Chairpersons.

5.     The committee will be composed as follows:

Drivers – Worker/Union (4)
Shop personnel – Worker/Union (1)
Office personnel – Worker/Union (1)
District Manager- Management
Route Manager - Management (2)
Maintenance Supervisor - Management
Safety Manager (Ex Officio)

6.     The secretary for the committee shall be provided by the employer Co-Chair, or be the chairperson not chairing the meeting.

7.     The employer Co-Chair shall be assigned by the employer. The employee Co-Chair shall be selected by the employee committee members.

# Appendix 1

## DUTIES OF MEMBERS

8. All members of the committee shall:

    (a) attend all committee meetings when present in the workplace for such meetings. Members may send alternates provided that such alternates are approved by the committee and by the union in the case of employee representatives;

    (b) participate in committee business, inspections, and investigations as required;

    (c) act immediately to resolve hazards by notifying the area supervisor, and a Co- Chair when appropriate; and

    (d) perform any other specific or implied duties assigned by the Occupational Health and Safety Act or the committee.

9. The committee secretary shall have the following duties:

    a) Report on the status of committee recommendations;

    b) Prepare the minutes;

    c) Distribute the minutes after approval;

    d) Disseminate information to committee members as required; and

    e) Assist the Chairperson as required.

10. The presiding Chairperson's duties shall include:

    a) Scheduling meetings;

    b) Preparing an agenda;

    c) Inviting specialists or other guests as required;

    d) Presiding over the meeting and guiding it as per the agenda;

    e) Ensuring a decision is reached on all agenda items;

    f) Ensuring deferred agenda items are assigned to a committee member;

    g) Reviewing and approving the minutes;

    h) Assigning projects to members; and

    i)  Ensuring the Committee carries out its functions and meets its obligations.

11.    Additional duties of the worker members designated by the committee may be:

    a)  Participate in workplace inspections;
    b)  Participate in investigations of fatalities or critical injuries;
    c)  Be present as required during any testing conducted in the workplace;
    d)  Accompany a regulatory inspector in the workplace as required; and
    e)  Be present during an employer's investigation of work refusals.

## MEETINGS

12.    The committee shall meet a minimum of nine times per year, during working hours, as agreed upon by the committee annually.

13.    Each meeting shall have a quorum. A quorum shall be the majority of the members present with at least half of those members being employee representatives.

14.    Additional or special meetings may be convened by either chairperson.

15.    Chairpersons shall alternately chair meetings on a rotation agreed upon by the committee.

## MINUTES AND AGENDAS

16.    The committee secretary, with the assistance of the chairperson, should:

    a)  Prepare an agenda prior to each meeting;

b) circulate and finalize the agenda prior to each meeting and make the time and place of the meeting known to the employees represented by the committee at least one week prior to the meeting; and

c) publish, retain, circulate, and make available minutes of each meeting.

## TERMS AND REPLACEMENT OF MEMBERS

17.   The normal term for a committee member shall not normally exceed two years. Members may renew their terms, or be reaffirmed by their union, as often as they wish. Members may not simply leave the committee without a replacement.

18.   Union/worker members wishing to leave the committee must indicate their wishes to the worker Co-Chair. The worker Co-Chair would ask the union to appoint a new member or arrange for a new member to be elected or volunteer. Where there is a union, the union must be informed, in writing, of any union position that becomes vacant by the worker Co-Chair. Replacements should be appointed as soon as possible.

19.   Management members wishing to leave the committee must inform the management Co-Chair who may appoint, or request that the district manager appoint a replacement.


Terms of reference adopted __day of, 20__


_____        _____

Management Co-Chairperson        Worker Co-Chairperson

## Appendix 2 Sample Simplified HSE Council/Policy Committee Terms of Reference

### Council Purpose

The Health, Safety, and Environment (HS&E) Council is a standing executive committee that is responsible for the overall policy and direction of Company HS&E issues.

### Scope & Mandate

The council shall represent the collective interest of all company employees throughout the company with regard to HS&E.

The council shall act in a decision-making capacity with full authority to make decisions or establish policy on behalf of the organization.

### Functions of Council

The HS&E Council shall:

- Determine corporate strategy, policy, and strategic plans regarding HS&E matters;

- Review and set goals, targets and objectives related to HS&E;

- Review past and present performance and determine areas of future effort and focus;

- Review and monitor the overall performance of systems set in place to protect the health and safety of employees, customers, contractors and the communities in which we operate and the environment;

- Consider and approve new initiatives and overall resourcing;

- Discuss and resolve issues affecting HS&E on a corporate level;

- Participate to the extent that it considers necessary in inquiries, investigations, and studies pertaining to HS&E;
- Review and address items forwarded by the HS&E Network; and
- Monitor the status of the company HS&E Management Systems.

**Composition of the Council**

The Council shall be chaired by the VP HSE.

The council shall be comprised of:

- President and CEO
- Vice Presidents of HR & HSE, Operations and Supply Chain Management
- Operations Managers of all Divisions
- Director of Equipment Maintenance
- Manager, Corporate HS&E

As required, sub-committees shall be assembled and mandated by the HS&E Council and shall be selected for their operational or subject matter expertise. Sub-committees shall report back to the Council.

**Member Responsibilities**

Council members shall:

- Attend all committee meetings (and/or arrange for an alternate or teleconference attendance if you are not available to attend);

- Identify agenda items to the chairperson;

- Review written materials prior to meetings;

- The committee members shall monitor the status of actions from the Council; and

- Attend semi-annual Health and Safety Summits (and/or arrange for an alternate if you are not available to attend).

The Chairperson shall:

- Preside over meetings as a standing member and chair of the council;

- Ensure development of meeting agendas;

- Invite others as required;

- Monitor meeting action items; and

- Communicate critical issues to the HS&E Council, as required.

**Meetings**

- The council shall meet a minimum of four times per year with attendance by the member or their designate being mandatory;

- Additional or special meetings may be convened from time to time to address critical issues

Terms of reference adopted _____ day of _____, 20__ .

_____      _____      _____

Chairperson          Position                    Signature

# Appendix 3

## Appendix 3 Sample format for minutes

## Sample #1

**Minutes of the Support Services Safety Committee. Meeting held 6 June 2018.**

Chaired by Linda Suffolk in the support services boardroom

**MEMBERS**

**Present**

Carl Ridley – Co-Chairperson. Management rep. Maintenance
Linda Suffolk – Co-Chairperson. Labor rep. Administration
Bob Brown – Labor rep. Human resources
Abdullah Sumarlin – Management rep. Finance
Valerie Instance – Management rep. Human resources
Larry Larue – Management rep.
Shelly Lu – Labor rep. Maintenance
Tom Sinclair – Labour Rep IT
Nancy Polenta – Ex Officio Safety Dept
Amanda Symak - Guest
Aaron Ramsay – Secretary

**Absent**

Mark Mensa (regrets). Labor rep. Cleaning staff.

| # | Item | Sponsor | Status | Target date |
|---|------|---------|--------|-------------|
| 1 | Call to order. Meeting was called to order at 10:00. | Chair | | |
| 2 | Review and approve agenda | Chair | | |
| | Review minutes of previous meeting | Chair | | |
| **3** | **Old business** | | | |
| 3a | Outstanding inspections items from cafeteria. Waiting for quote on repairs. | Bob Brown | In progress | 21 Jun 2018 |
| 3b | Follow up on corrective actions incident 18-038. Larry signed off on the actions and marked them as complete. | Larry LaRue | Completed | 6 June 2018 |
| 3c | Review of Respiratory Protection Program for facility maintenance staff. | Shelley Lu | Open | 21 July 2018 |
| **4** | **New business** | | | |
| 4a | Presentation on working alone challenges for outside workers. (see attachment) | Amanda Symak | | |
| 4b | Quarterly incident review | Safety Dept | | |
| **5** | **Business from the floor** | Chair | | |

# Appendix 3

| # | Item | Sponsor | Status | Target date |
|---|------|---------|--------|-------------|
| 5a | Bob Brown indicated that some employees were unsure on how to report safety problems. Bob will review the employee orientation and follow up with safety rep | Bob Brown | Open | 15 Sept 2018 |
| 5b | Shelly Lu suggested that the committee get a product rep in to give an information session to the committee on respirators | Chair | Open | 15 Sept 2018 |
| 6 | **Adjournment** | Chair | | |
| 6a | Next meeting July 10th, 2018, in the support services boardroom. Chaired by Carl Ridley | Chair | | |
| | Meeting adjourned at 10:46 | | | |

APPROVED

_____          _____

Carl Ridley                                     Linda Suffolk
Management Co-chair                              Labor Co-chair

## Minutes Explained

Minutes are the primary method by which the committee records its activities and communicate with the workforce. These must have a simple structure but communicated a great deal of information.

Looking at the first example provided for minutes, the following explanation is provided

**Name of the committee, meeting date and time** - it is not always intuitive to include this information, but it is important. At the start of all minutes, and ideally in the header of the document, the name of the committee should appear along with the meeting dates and time. This allows anyone reading the minutes to clearly see the date and time of the meeting and when the next meeting might be.

In the minutes, it is important to identify the location and time where the meeting took place. Most committees invite employees to come to meetings as observers or to speak to the committee on a matter of concern.

The person who chaired the meeting should also be identified in the minutes. This is to demonstrate that the chairpersons take turns chairing the meetings. It also communicates who chaired the meeting to whoever is reading the minutes.

**Members. Those present and absent** – At the start of the minutes, there should be a listing of those who were present at the meeting and those who are absent. Members of the committee should be identified by noting if they are a management or union/labor representative, and what area(s) they represent.

Those present in an unofficial capacity such as a member of the Safety Department may be noted as being ex officio. Speakers, or other people giving presentations, can simply be noted as a guest.

This part of the minutes is important because it allows people to see who is representing them at the committee in case they wanted to speak with that person. It also shows them who was present at the meeting.

**Item #** - You may want to number the items on your agenda. Any numbering scheme you wish is acceptable. Some minutes use Roman numerals (as seen in the agenda example), and others number things sequentially.

In this example, the main headings are numbered, and the items below them are assigned their own number. This is for ease of reference so that anything numbered "3" is known to be old business. Another option is to use headings and not number the items at all simply.

**Item** - The column identifying the item can serve a single or double purpose. In some cases, minutes may separate the title of the item from the discussion at on its status. In this example, the discussion around the item is contained in the same column. This is often done to save space as minutes that are produced in a landscape format tend to be much more difficult for people to read.

**Sponsor** The sponsor is the person responsible for the item. This column is often titled "responsible" or "accountable." This is important to establish who is responsible for taking action, or leading action, on that item. In the example, we see various members of the committee being the sponsor for specific things within the minutes.

**Status** – The status of an item must be tracked in the minutes for every single meeting. Normally the status is limited to open, closed, pending, or in progress.

No item should be dropped from old business, or new business until it has been marked as completed. This allows people reading the minutes to track an item through to completion once it has been identified by the committee.

**Target date** – This column may also be termed "due date." This column is normally included to indicate the time frame in which the committee feels they must resolve the issue. Normally target dates are not more than ninety days. Target dates can be revised from time to time if the committee needs to get additional information or has other priorities.

# Appendix 3

## Sample #2

**Workplace(s) represented:** Support Services

**Number of workers in workplace:** 48

**Date of meeting:** 6 June 2018

**Date of next meeting:** July 10th, 2018

**Meeting Chaired by:** Linda Suffolk

| Employer Cochair | Carl Ridley | | | | Worker Cochair | Linda Suffolk | | |
|---|---|---|---|---|---|---|---|---|
| **Management members** | **Dept** | **Present** | **Absent** | | **Worker members** | **Dept** | **Present** | **Absent** |
| Carl Ridley | Maint | X | | | Linda Suffolk | Admin | X | |
| Abdullah Sumarlin | Fin | X | | | Bob Brown | HR | X | |
| Valerie Instance | HR | X | | | Shelly Lu | Maint | X | |
| Larry Larue | Maint | X | | | Mark Mensa | Jan | | X |
| | | | | | Tom Sinclair | IT | X | |

| Item | Issue | Action | Target date |
|---|---|---|---|
| 1 | Meeting called to order at 10:00. Agenda reviewed and approved. | | |
| 2 | Outstanding inspections items from cafeteria. | Waiting for quote on repairs. | 21 Jun 2018 |
| 3 | Follow up on corrective actions incident 18-038. | Larry signed off on. Marked as complete. | Complete |
| 4 | Review of Respiratory Protection Program for facility maintenance staff. | In progress | 21 Jul 2018 |
| 5 | Presentation on working alone challenges for outside workers. (see attachment) | Info | N/A |
| 6 | Quarterly incident review | No action identified | N/A |
| 7 | Bob Brown indicated that some employees were unsure on how to report safety problems. | Bob will review the orientation and follow up with safety rep | 15 Sept 2018 |
| 8 | Shelly Lu suggested that the committee get a product rep to give an info session on respirators | Linda Suffolk will work with Shelly to identify product rep. | 15 Sept 2018 |
| 9 | Meeting adjourned at 10:46 | | |

Minutes are an accurate representation of meeting.

_____                    _____

Employer Co-chairperson                         Worker Co-chairperson